I0407593

Table of Contents

A Worm Farm at Your Own Home

DIY Worm Farming (Vermiculture)

Vermiculture need not be carried out on a commercial scale, but as a very useful and interesting home project that can involve the whole family. For the moment, let us confine ourselves to investigating the benefits of DIY vermicomposting as a special hobby or even as a small scale commercial enterprise carried out from one's own home and postpone any discussion of large scale commercial vermiculture farming until later. Don't worry, your venture into worm farming is not going to be like opening your home to some sort of smelly Frankenstein monster. Worm composting is a very wholesome process; handled correctly, it does not create bad odours and moreover it gets rid of nasty wastes without adding to the burden of global warming.

Setting Up a Worm Farm with Your Children

As we have already said, vermiculture can involve the whole family – the educational benefit of sharing a worm farm with young children is obvious, especially if combined with an organic vegetable garden. High yields of healthy fresh vegetables and fruit can be produced in a very small garden from the rich fertilizer produced by your worms. We all know that it's a battle to get kids to eat healthy vegetables, but children who grow their own crops will always delight in eating them. What an investment in their future health! But equally importantly – worm farming itself is great fun!

How to Get Started with Vermiculture

As a potential worm farmer your first choice would be to decide, whether to make your own worm farm or to simply buy one of many convenient systems, readily available on the market. Ready made worm composters are usually supplied together with a starter pack of worms and also with operating instructions. These fancy "designer" composting kits can be obtained directly from manufacturers, ordered through the internet, or found at many garden centers, plant nurseries or similar outlets. However, although convenient and very user friendly – as specially created products, they are certainly not all that cheap .

Building Your Own Worm Farm

A more interesting and cost effective alternative would be to consider the challenge of building your own worm farm from scratch. Making a DIY worm composter (often miscalled an earthworm farm) is not rocket science and it is quite easy to assemble one to fit your own requirements. Your worm farm could be constructed from timber, brick or metal, but it is far easier these days, to adapt general purpose plastic bins from the local supermarket or hardware store for this purpose. Other web pages on this site give step by step instructions for making both a simple single bin composter or the neat three bin worm farm shown above. It is not at all difficult and making your own worm farm this way would save you money that could be used to buy plants or seeds for your garden!.

Money from Farming Worms

The Direct Way

Once an established worm farmer, you might want to expand your vermicomposting operation into a full scale home industry, in order to profit from all your efforts. With a little imagination, profit can be derived either directly or indirectly from the worm farming operation itself. Worm farming is unlikely to make you very rich, but with a little entrepreneurial vision and some aptitude for marketing, vermiculture can definitely save you money and could also provide a useful steady income for you. The direct approach, is easy to understand. You would be simply have to find a way to sell the worm compost and worm tea from your vermiculture operation to any nearby gardeners who understand the benefits of organic gardening. Selling organic potting soil is another option, (4 vermicompost/ 4 river sand/ 4 stiff clay plus a half measure of bone meal). At the same time your excess compost worms would be sold to aspirant "Newbie" worm farmers in your neighborhood, or otherwise made available to fishermen, who always value a supply of red wigglers.

The Indirect Way

A more commercial approach, with far greater potential profit, would be to contact worm farm suppliers or manufacturers directly, and ask to become a reseller of their branded worm farm kits. Most manufacturers of the popular brands of worm composting kits will sell their units and a range of vermiculture accessories, at wholesale prices to resellers. You could simply act as a distributor and sell them on for a profit. However, a much better proposition would be to "add value" by setting the kits up as "ready-to-go" worm farms, complete with bedding, a couple of hundred worms, a few accessories and a set of typed instructions. And for those handy with tools, the optimum strategy would be to substitute homemade DIY worm farm kits (such as those described later) for the commercial branded units – then set them up with your own worms, and sell them as "starter kits". As they wouldn't cost too much to set up, these DIY units could be sold at a good discount to the fancy branded kits, to make them attractive. On the worm farming side, you would mainly be interested in breeding and harvesting the worms, in order to populate your kits, and thus would need a fairly sizeable operation. Obviously the by-products – vermicompost and vermitea, would also be sold or ploughed into the garden to raise saleable produce.

Marketing

As with most new enterprises, the biggest challenge would be probably be the marketing. Ideally you could try persuading local plant nurseries to sell your units or products, either directly or on consignment and then combine this approach with advertisements on community notice boards, or by way of the internet or through the local papers. Many vermiculture entrepreneurs run a tables at local fairs or flea markets, demonstrating the process with an active worm kit and using the opportunity to take orders for worm farms and worms, while directly selling vermicompost and worm tea to the general public. Make no mistake, it will never be easy to get started. But have a little patience – establish a catchy brand for yourself, sell a fair product at a fair price and once your name gets around, things will definitely become easier.

The Old Fashioned Way

If the foregoing seems too challenging for you, just remember that there is an alternative way to make composting worms work to the benefit of your pocket. By starting your own organic garden, fertilized by worm castings and liquid worm tea, you would not only be improving the integrity of what you are eating, but you would also significantly reduce your supermarket vegetable purchases as well. At the end of the month you will have more money in your pocket. And then, if you have the energy (and space), you might even have excess garden produce to sell to your neighbors.

What About the Workers? – Earthworm Versus the Redworm

In reality all these worms are technically species of earthworm and have much in common physiologically. However common usage assigns the name of earthworm to the deep burrowing garden worms that inhabit the soil around the roots of plants, whilst reserving the name of compost or manure worms to the varieties that are found in dung heaps, fallen leaves or other plant trash, right at the surface.

Improved Soil Fertility Through Worm Activity

The world would be a sad, dirtier and hungrier place were it not for the humble garden earthworm and his/her* useful cousins known collectively as compost or manure worms. The red manure worms (technically known as epigeic or detritivorous earthworms) are actually the main heroes of this web site – but homage must first be paid to the basic hard working garden earthworm, for Charles Darwin, the famous 19th century evolutionist, who wrote Origin of the Species, said of them : –

"It may be doubted whether there are many other animals which have played so important a part in the history of the world, as have these lowly organised creatures."

*Although his/her seems very PC – the real point is that all these worms are hermaphrodites, having both male and female sex organs.

The Common Earthworm

Actually several worm varieties have always been busily hard at work, striving to promote soil fertility – and this was long before a certain species of naked ape left the trees and turned his attention towards the invention of the plough. Common earthworms such as lumbricus terrestris, are probably the most efficient biological agents to be found anywhere in the world. They specialise in removing dead organic material from the surface of the land, greatly enriching it in the process, and then the clever earthworms carry the improved residue deep underground, right down amongst the roots of plants, where it is most needed. The hard working earthworms aerate and loosen heavy soils, improve water retention and simultaneously enrich the soil's fertility with their faeces, which we call worm castings.

Our Hero – The Red Worm

However, our own particular heroes – the more aggressive red manure worms, always remain near the surface , where they are to be found partying in heaps of animal dung, or wriggling around in layers of decaying leaf mould and other interesting plant detritus. Because they live at the surface they are "epigeic" (gk = above the earth) and as they eat plant detritus, they are "detritivorous".

 These highly active red worm colonies speed up the entire natural composting process, by many months, by literally chewing through heaps of dead organic material, whilst continually beneficiating the fertility of the resultant humus with the richness of their castings, which are eventually utilized enthusiastically by plants.

Worm castings are an extremely good plant food as they are always rich in nutriments, minerals, beneficial microbes, enzymes and plant hormones. Beneficial microbes associated with vermicasts have been scientifically shown to directly reduce bacteriological pathogens in soils upon which the worm compost is spread.

The Role of Microbes in Vermiculture

The relationship between earthworms (including the various composting worms) and the aerobic microbes or bacteria that accompany them is one of nature's most perfect examples of symbiosis. The worms have millions of beneficial bacteria associated with them, both externally, on their skin, in the mucus secretions that keep them moist and also swarming internally inside their gut. Worms have no teeth, bills or jaws, nor a true stomach and rely on the bacteria swarming around them to actually break down the foodstuff that we put in our bins. The deconstituted foodstuff is altered considerably, such that it can be sucked up by the worms as a slimy paste-like substance. It goes directly into their gizzard and passed onward through a very rudimentary digestive tract, together with the masses of bacteria that are swarming within the slime. Inside the worm's gut the breakdown process continues and the worms' digestive tract, provides a perfect environment for the ingested bacteria, who multiply further and continue to convert the complex cell structure of the original foodstuff into its basic elements and compounds, altering it into a simpler form that can be used directly by both the worms and the bacteria for nourishment. These simple elements and compounds provide the basic building blocks to sustain both worms and bacteria and are reconstituted according to the messages carried by the DNA to build up the complex cell structures that create the living physiology of both worm and bacterium. A true win / win situation for both organisms. Large numbers of these bacteria are released back into the worm bin, together with the waste products in the faeces or castings – our vermicompost. The microbes will have multiplied in the ideal environment of the worm's gut and now, greatly increased in numbers, are once again ready to attack new food sources and start the process all over. Of great importance, these waste products, or vermicompost, excreted by the worms have been thoroughly processed by the microbes and are now in the form of simple elements and compounds, that are readily taken up by our garden plants, providing a highly nutritious food for them. Moreover any dangerous toxins and infected material would have been simultaneously neutralised by the bacteria within the worms gut, as complex forms of pathogenic material are also broken down into simpler, more basic (harmless) components by the microbes. In the soil the process continues and worm compost, with its load of beneficial bacteria will also tend to improve the health of soil around the roots of plants by removing pathogens. This is the beauty of using worms and their huge army of tiny microscopic helpers, for your composting.

Suitable Varieties of Composting Worms for Vermiculture

For the worm farmer, wanting to set up worm composting bins, the red manure worms are a far better bet than the more stolid greyish-brown earthworms. The lively reds, reproduce far more quickly and are much easier to manage, because their habitat is epigeic i.e. at the interface with the surface. Unfortunately the common names of the different species of composting worms are confused by loose terminology and sometimes different worms are called by the same name. Unless the scientific (latin) name is also used, there is likely to be some confusion. The most common manure worm used in worm farming in the US is the red worm, (Eisenia foetida or fetida), alias redworm, red wiggler, red wriggler, brandling worm and often confused with the similar looking tiger worms (Eisenia Andrei). In the UK the larger nightcrawlers (dendrobeana) are much favoured for worm farming, especially for fishing worms. A species of European worm, the driftworm, also known as Red wriggler (Lumbricus rubellus) is also commonly used in vermiculture, especially for fishing bait as it is large, lively, robust and is even suitable for salt water fishing.

Latin Names for Worm Species

Because there are so many overlapping common names in use, such as red worm, redworm, red wiggler, red wriggler – the safest way to know you have the right worms is to use the scientific (Latin) name to identify the species.

Eisenia andreia:

Usually called the Tiger Worm, because of alternate bands of darker and lighter red colour. Often confused with Eisenia Fetida (Foetida) and to make things worse they are also known as Red Worms. Like Fetidae They are quick breeders and productive in vermicomposting and good fishing worms. They are between 2 to 3 inches long and weigh in at 900 to 1000 worms per pound. They are found throughout the world and as such are no threat to the environment if they escape.
Temp range – Extremes: 38ºF-88ºF/Optimum 70ºF -80ºF

Eisenia fetida (foetida):

Commonest compost worm used in worm farming and easy to obtain. Usually called Red Wigglers, but also known as Red Worms, Red Wrigglers, Compost Worms, Manure Worms and Brandling Worms. They got their name of red wiggler because as fishing worms as they are active on the hook and stay alive in water for some time, although they are a bit small for this purpose. They are between 2 to 3 inches long and weigh in at 900 to 1000 worms per pound. They are quick breeders and productive in vermicomposting. They are found throughout the world and as such are no threat to the environment if they escape. Temp range – Extremes: 38ºF-88ºF / Optimum 70ºF -80ºF.

Eisenia hortensis:

Common name: European Nightcrawler also commonly called Redworm, it is much bigger than Eisenia Fetida (foetida). It is a quick breeder and a good composter (makes plenty of castings). Much sought after for fishing bait, as it can tolerate near freezing water and is one of the few "earthworms" suitable for salt water fishing. These worms can grow up to 7 inches in length, but usually are between 3 to 4 inches. 300 to 400 worms per pound.

Eudrilus eugeniae:

Common name: African Nightcrawlers. These worms are much larger than Eisenia Fetida (Red Wigglers) and are commonly over six inches long. Good compost worms and great for fishing, because of their size and as they are lively on the hook and have a firm skin. They prefer temperatures of around 75ºF- 85ºF , but can tolerate 45ºF- 90ºF, cannot tolerate extreme cold and dislike disruption of environment and handling. Weight: 175 to 200 worms per pound.

Lumbricus rubellus:

A species of European worm, the driftworm, also known as Red wriggler. It is actually an burrowing earthworm and not a true compost worm, but in nature is Endogeic and feeds close to the surface. It is a large worm of average length 4 inches and is commonly used in vermiculture, as it is very productive at cooler temperatures. The optimum temperature is around 50ºF and it only stops breeding around 40º. Rubellus is also attractive as a bait worm as it is large, lively, robust and is even suitable for salt water fishing. However there is real concern that Lumbricus rubellus, as an exotic, could become a problem invasive species in North America and there are claims that it is spreading into the northern woods and causing damage to native forests. This is because it tolerates lower temperatures and wetter conditions than most compost worms. It causes damage by breaking down the plant subterranean trash that protects the surface roots of trees. Because it can burrow deeply, it can overwinter when the surface becomes frozen, unlike most compost worms such as Eisenia fetida. So before you start your worm composting – it is important that you check local requirements and choose the right worms for your area and never throw unused bait into the forest.

Lumbricus terrestris:

Common earthworm species, sometimes called nightcrawlers . They are not suitable for vermiculture as they are a deep burrowing species (Anecic). Their burrows, are semi permanent and may extend to six feet below the surface – these burrows are lined with mucus and help aerate the soil and improve water retention.

Perionyx excavatus:

Common name : Indian blue worm. This species has a distinctive iridescent blue sheen to its skin. It is a tropical worm and does not tolerate cold or much handling or environmental disruption.

Although small, it is suitable for vermiculture as it is a prolific breeder and matures quickly. It has one major drawback though – it is known for staging mass escapes from the worm farm, for no apparent reason and is somewhat unpopular for this reason. Temperature range – Extremes: 45ºF – 90ºF / Optimum 70ºF – 80ºF.

Mix and Match Your Worms ?

Many worm farmers prefer using a mix of Eisena fetida or foetida together with Eisena Andrei . Some composters claim that yields are increased further by adding the European Red Wriggler, Lumbricus rubellus to the menagerie. Each species has different requirements as far as temperature preference and growing conditions and would produce better or worse in different situations – hence the advantage of setting up cocktails of different species. Fortunately hybridization does not seem to be a problem.

However there is some concern that the large red European Lumbricus rubellus is an invasive species in North America and there are claims that it is spreading into the woods and damaging the native forest. So before you start your worm composting – it is important that you check local requirements and choose the right worms for your area. Just remember, even if you try to separate live worms from the worm castings, you will inevitably have some egg casings left in the vermicompost that you spread on your garden.

Worm Reproduction

All species of earthworms, including the worm farmer's favorite, the red wiggler (Eisenia fetidae) and the common earthworm (Lumbricus terrestris), reproduce in much the same way. Earthworms, according to Charles Darwin are "lowly organized creatures", but it must be said that to most of us, their reproductive system seems highly complex and rather weird.

Hermaphrodites

The first thing to understand is that all earthworms are hermaphrodite. This means that each individual has both male and female genitalia and therefore cannot be considered exclusively male nor female. When two worms couple for mating, both worms produce sperm, which is exchanged directly with their partner during the sexual encounter and the exchanged sperm is subsequently used to fertilize the eggs of each individual.

Fertilisation

The various worm species take different times to reach sexual maturity. For Eisenia fetidae and other composting worms it is about eight to ten weeks. The sign of approaching maturity is the development and increasing prominence of the clitellum band. This is a swollen ring or belt of lighter coloured flesh that develops about a centimeter or so behind the mouth at tip of the worm. This bulge is usually a light pink color, but in some species the clitellum has a distinctive yellowish tinge. The clitellum band has a leading role to play in the reproductive cycle of the worms. When the mature worms begin ovulating they seek out sexual partners and begin to twist around each other either in pairs or more often in small tangled groups or clumps. As the sexually aroused worms entwine with each other, their bodies are stimulated by the mating process and large quantity of sticky sperm is produced, which attaches itself to the upper bodies of their partners. Shortly after the coupling has been successfully completed, the clitellum band begins to excrete a mucous substance that soon stiffens into a jelly-like ring. This donut shaped ring then gradually loosens up and begins to slip off the worm. As it passes over the worm's body the jelly ring collects some of the attached sperm and at the same time also gathers up a number of eggs that are simultaneously released in its path. As it slips off the "parent" worm, the jelly ring curls itself into a ball , enclosing some of the eggs and sperm within itself in a protective cocoon. The outer portion of the cocoon quickly hardens into a shell or casing enclosing the inner gel mass, which provides a nutritious environment for the fertilization and gestation of the eggs.

Casings

The cocoons are about the size of grape pips and have the shape of birds eggs. They start off at a kind of yellowish grey color but in time become more of a mahogany brown. These cocoons or egg casings, as they are often called, provide a safe environment to protect the fertilized embryos. The cocoon is the equivalent of an external womb, keeping the the embryos safe until they are fully developed. The baby worms emerge from the cocoon around seven weeks after fertilization of the eggs and each egg casing would typically give birth to between seven and fifteen pale pink baby worms.

Traditional Worm Farming

For all its current popularity worm farming is nothing new. "Vermiculure" and "vermicomposting" have the kind of sophisticated pseudo-scientific ring that leads one to assume we are dealing with cutting edge technology, run by serious minded people in white lab coats. But the plain truth is that people with very dirty hands have been making their own worm farms for centuries. They loosely called them earthworm farms or fishing worm farms and the object was simply to maintain a good supply of fishing worms for the angler and his friends throughout the year. As the worm compost aspect was of secondary importance to the fisherman, who was really only interested in growing worms, the general arrangement could be pretty crude, but it was usually effective.

A Basic Method of Raising Fishing Worms

A simple fisherman's worm farm, consisting of a single compost bin, was traditionally made from a wooden crate or a rectangular brick, concrete or steel trough. It would have a perforated base or slots for drainage. Nowadays, the bin would usually be made from opaque plastic, such as the one portrayed below, with a lid and would be set up on a few bricks, to assist drainage. It must be kept out of direct sunlight to avoid excessive summer heat and provision would also have to be made to protect the earthworm farm from the winter cold, in areas subject to frosts. In the traditional method, a thin drainage layer of gravel or sand would be put in to cover the holes/ slots in the base and then a bedding of well rotted manure or old compost, mixed thoroughly with wood shavings or shredded newspaper, would be layered down above this to receive a few hundred worms. Enough water would be sprinkled around to keep everything damp, but not saturated. The mix would then be covered with further bedding material and a sheet of damp burlap (also known as sackcloth or hessian) would be draped over the worm farm, to help keep temperatures even and to keep out flying pests.

a) The bin is of opaque plastic, with a lid, it is at least 8 inches (200mm) deep

b) Drill 1/4 inch (6mm) holes in bottom and sides for air and drainage

c) Put the bin on blocks to allow for drainage and set it up in protected place in shade

Care of the Worms

A few days after the worms have been introduced, the fisherman/ worm farmer should begin to add organic material to feed them. It is important not to add too much food material at any one time, (to any kind of worm farm) – and especially not fresh manure. Besides attracting pests such as mites and springtails, a large biomass will hasten bacterial activity, which can result in a fast exothermic reaction amongst the decomposing organic material, producing high temperatures. This exothermic reaction is normal (and essential) for ordinary garden composting, but can be fatal in worm composting, as the excessive heat build up can easily kill off all the worms. Rather allow some pre-composting of fresh manure before adding it to the worm bin.

Once the worms have become active and started to produce compost, the farmer should stir or rake up the food/ compost mixture every few days (with a fork or with his hand). This is to aerate the composting mix. Worms love the dark, but need plenty of oxygen. For this reason you should use an opaque bin – never a transparent or translucent one and always provide sufficient air holes. After the raking, add in more food for the worms and cover again with the bedding material and replace the lid – this helps to help keep pests at bay. Never let the bedding become completely dried out – generally there should be enough moisture in normal kitchen waste to keep the compost moist, but don't hesitate to spray a little water around if you think it is necessary.

As long as there are no extremes of temperature and the bedding remains damp, but not saturated, the worms will flourish and proliferate on a regular supply of kitchen scraps. From our perspective, as vermicomposters,, the only problem with this simple single bin unit is that it is not easy to separate the worms from the compost. This fact was never important for the fisherman, who would simply dump both into his bait can . However a more efficient system is required, if we want to "harvest" the compost, while leaving the worms behind to continue the composting process.

Established Ways of Separating the Compost from the Worms

A part solution for this problem would be to make the worm farm in the form of a longish trough and then start the worm composting from just one end of the bin. The idea is to add the worm's food progressively in stages, moving along the length of the trough – thus feeding the compost worms only at the leading end, and so encouraging them to slowly make their way along the trough, away from the starting point, which would contain no further fresh food for them – The worms would thus slowly migrate along the trough, after the food, leaving behind them the already composted worm castings – the very product we seek. When they reach the end, simply start the process in the opposite direction, after removing the all the usable compost, which by then should hopefully be more or less free of worms.

The Stacked Tire Worm Farm

If space is a problem, much the same process as described above can be carried out vertically, by progressively building up a stack of used car tires on a well drained base. Waste material, as food, would be added from the top, which would be covered to keep out pests and the stack would be raised, by adding a new tire whenever required. The worms would be most active in the upper feeding layer and migrate upwards towards source, while the worm compost would then be collected by pulling the bottom tire horizontally out of the stack, which with a little effort can be done without disturbing too many worms. You'd have to be strong for this, or use a bar for leverage, as car tires full of damp worm castings are not light!

You set up the first tire on a perforated board, placed on a few bricks. This is for drainage and will also allow for the upward flow of air through the stack.. Better still, if you wish to collect the worm tea, the tire can be set onto a slightly sloping corrugated metal roof sheet, which will direct the liquid run off into a container. Initially you are now going to treat this first tire as a single bin composter, as described above. However, when the worms are well established and the compost has more or less filled the first tire, add a second tire above the first. It is less messy in the long run if, at this point, you separate the two layers with burlap or even a well perforated plastic sheet, which will allow the upward migration of the worms and some ventilation. To improve the air flow, you can place a few thin slats or rods to open a few small air gaps between the two tires. After this you just continue as before, adding tires above, as you pull out the compost below.

The beauty of this system is that it costs next to nothing and it is easy to expand the scale of the operation, by bringing in more old tires to put up multiple banks of individual worm farms – as many as desired. Admittedly, the stacks of tires would be a bit unsightly in a suburban home, but this system would be an excellent addition to any school feeding garden in a poor community.

How to Make Your Own Worm Farm

Why buy an expensive worm farm, when you can set up a perfectly good stacking system wormery, for less than half the price of buying in a fancy branded worm farm from a dealer? You won't even need to be much of a handyman, nor use expensive materials to produce a neat unit that will look good and function well.

The Principle of the Stacked Bin Worm Farm

Traditional methods of vermiculture have their place, but today's suburban worm farmer wants a composting system that takes up minimal space, looks good and is clean and convenient to use. The home worm farmer, or amateur vermiculturalist can use suitable modern products and a better understanding of the habits and requirements of the compost worms in the worm bins to design a system that is both convenient to handle and efficient in the usage of materials and manpower. The principle of the stacked bin worm composter is that, unlike the drab earthworms, who dig deep, our red compost worms always migrate upwards, towards the food, leaving their castings to fall below them. We use this information about red worms to our advantage. Generally the idea is to build up a multiple stacking system of connected worm bins or trays that are slightly tapered to allow the bins to nest, one within the other. Worm castings (the compost) are collected in the lower bins and worm food (kitchen or garden scraps) is consumed in the upper levels of the wormery. When a lower bin is nearly full of castings it is emptied and rotated to the top and so on.

Choosing Your Bins

The size and number of the nesting bins is variable, depending on the desired scale of the operation. Common plastic storage bins, sold for general household use at hardware stores, supermarkets and camping goods outlets are quite suitable for making your worm farm. Usually the sides are not vertical, but slightly tapered for convenient stacking on the retailer's shelves – this suits us, as it allows for partial nesting of bins . A lid would be required for the top bin. Worms hate light – so don't get opaque bins. Heavy black bins are good. The plastic storage containers are not expensive and come in a variety of sizes. For a small scale composting set-up, for processing kitchen waste, three containers of about 45 litre (ten gallon) each would be adequate. For processing a greater amount of waste such as from large gardens or stables, bigger bins with more tiers can be set up, just as easily.

Instructions for Creating Your DIY Worm Composter

The Sump

The lower sump bin is configured differently from the upper bins and would be prepared first. Its function is to collect excess fluid leachate, called worm tea, or compost tea.

The sump may be fitted with a 3/8 inch (15mm) barrel tap, through a small hole drilled in the base for conveniently draining out the excess fluid (the worm tea) that will accumulate there. This tap is not essential, but would avoid the otherwise potentially messy job of having to tip the worm tea out by rotating the bin.

If you do decide to put in the tap, make sure it seals well in the hole, by providing good washers and lock nuts.

The Composting Bins

The two upper bins will actually hold the worms. They are to be identical and are prepared as follows : –

Drill a pattern of ¼ inch (6mm) holes across the entire base of each container for drainage and to allow for ventilation and the upward migration of the compost worms, these holes should be regularly spaced at approx two inches (50mm) centres apart in either direction.

For further aeration, drill a row of ¼ inch (6mm) holes at two inch (50mm) centres, in a continuous line around the walls of each of the bins. This line of holes would be about four inches (100mm) below the top rim of the bin.

It is not essential to drill holes in the lid, which is closed tightly over the upper bin. as you should get enough air through the sides.

Setting It Up

After preparing your bins, you first set up the lower (sump) bin on bricks or blocks, allowing enough space to tap off the fluid from beneath it. Choose a shady location for the worm farm (in a shed or garage, if you are subject to frosts).

The second and third bins are "nested" within each other and dropped into the sump bin. To maintain a working space for the worms, and for accumulation of compost, you need a few spacers or packers of about six to eight inches height, between the two upper bins and some smaller packers of about four inches in the lower (sump) bin. You can use wood blocks or sealed food jars for packers.

The packers also prevent the tapered worm bins from jamming together and cause a gap between the bins, which improves ventilation.

To prevent "nasty bugs" from squeezing in between the bins, you should close (caulk) the small gap between them with strips of shade cloth, or mosquito netting.

Starting Production

Now you are ready to go into production : –

Set up your worms in the top bin with a good (damp) fibrous bedding such as coconut coir, (or just shredded newspaper), put in a little compost and a handful or two of damp soil with the worms and after a few days you will be ready to start feeding in your kitchen scraps. Cover the food with more bedding material to discourage pests and keep the lid closed.

Make sure the worm farm is never allowed to dry out, by sprinkling water over the bedding periodically, if there is not already enough moisture coming from the food scraps.

When the top bin has been fully productive for a while, the worms will multiply and compost will be start accumulating from the worm castings. When the quantity of compost is meaningful, stop putting feed into this bin and swap over the upper two bins by putting bin No 2 to the top of the stack, with bin No 1 now in the middle.

Set up this new top bin with clean bedding, a small amount of the old castings and immediately start feeding your kitchen scraps into it. Over a few days, the worms will naturally migrate upwards towards the new food source, leaving the lower bin with only a few stragglers and it should be ready for the harvesting of your compost within about three weeks after the swap.

To get at any specific layer, to add food, bedding or to remove the vermicompost, just lift off all the overlying worm bins, one by one until the desired level is exposed for examination and then replace them in the same order. They will not be too heavy – but don't try lifting more than one layer at a time, unless you have a good chiropractor!

All you need to do is to keep repeating the process of alternating the top two bins on a regular basis, taking out the compost, whenever it accumulates, and tapping off the worm tea from time to time. This vermitea, is a very valuable product as it is a highly concentrated liquid fertilizer that can be diluted for immediate use on your garden.

Creating the Right Environment for Your Worms

Bedding

To set up the composting in a new worm bin, a fairly damp, but not saturated worm bedding layer, preferably of fibrous material such as coir, wood shavings (untreated) or carpet under felt, would be laid over the perforated base of the bin. Shredded newspaper can also be used. A thin layer of damp garden compost or well rotted manure would give the optimum temporary home for the worms, until the process is underway. Depending on the size of the bin, a couple of hundred red worms should be enough. Cover the worms with some more of the bedding material, to keep out flying pests and after a few days start adding food scraps under the top layer.

The Feeding

It is always better to mash up the kitchen scraps before feeding the worms, but an even better idea is to first place the scraps in a plastic bag in your freezer as freezing will greatly speed up the feeding process as it breaks up the cell structure of the worm food, making it easier for the worms to digest the material. Avoid citrus, pineapples and onions as these make the material too acid and too much meat and fat is not good, especially as these may attract rats – which can gnaw right through plastic. Every so often, replace the worm bedding with new material. Some soil or sand is also needed in the worm's diet, as they use this in their gullet to act as a grinding/ calcifying medium. Crushed eggshells or a little agricultural lime will raise the PH, if the composting environment becomes too acidic. A PH neutral environment is optimal.

Your Worms Need Air!

Your worm farm, should always smell good, and have a slight earthy odor. If you notice a sour/ rank smell developing and the bedding and compost is beginning to look over damp and slimy, with possibly some fungus present, it probably means that the environment is becoming anaerobic and primarily needs better ventilation to bring in more oxygen. This condition may be caused by excessive feeding, too much greasy food, such as meat and dairy, acidic conditions or not enough air circulating. Firstly, make sure the ventilation holes are not blocked and that the drainage is effective, then fluff up the bedding and rake up the vermicompost, to allow the air in. If there is too much unprocessed food lying around, stop feeding for a few days and thereafter put in less food, or get more worms. A we have said, it is also important to avoid putting in too much greasy food and acidic fruits such as citrus and pineapple.

Acceptable Temperature Range

Temperature is important and compost worms generally prosper best at temperatures that we would be comfortable with ourselves. Although different species have their preferences, as a generalization, they will breed at temperatures as low as 12º to 18ºC (54º to 65ºF) but stop all activity under 8ºC (46ºF)and as the temperature increases to around 25ºC (80ºF) they will become more active and productive, but above this the performance will drop off. As the temperatures rise or fall much outside this range they become more and more at risk and steps need to be provided to protect them from extremes, by having the worm farm set up in a shed or garage, where the temperature can be controlled. We need to be careful of not putting too much organic waste (especially fresh manure) into our worm farm at any point as the natural rotting process gives off heat and a bed of compost can easily get too hot for the compost worms to exist in it. The solution is to heap up this sort of material away from the worm composter for a few days to allow the heat to be generated and then dissipate naturally. Once it has finally cooled down it will be safe to use.

Handling Vermiculture Pests and Other Problems

Ok – you've started your worm farm for vermicompost, worm tea, worm castings, to do your part against global warming, to provide bait for fishing – or whatever. But suddenly things start going wrong! Pests can be a big problem. So lets roll up our sleeves and get our hands dirty as we look at some ways to prevent pests from ruining your worm bed.

Protecting Your Worm Bed (prevention is better than cure)

The best way to handle worm farm pests is to ensure that they don't establish themselves in the first place. Therefore it is best to keep your worm beds well maintained by ensuring that:

Your bin lid or farm enclosure is secure.

The worms and bedding are covered with either a sheet of plastic or a damp sheet of burlap (Hessian).

Food scraps are covered with bedding to prevent them becoming mouldy and attracting pests.

No meat, greasy food, or pet faeces is included in the feed as these attract flies – therefore maggots – and possibly even rats, which can literally gnaw their way into plastic bins.

For continuous worm farming, it is recommended that you house your worm bin, or other worm farming medium, in enclosed places such as: garages, sheds, basements or out-buildings; therefore making them less accessible to pests. It would also be helpful to screen the buildings as will help limit your losses to rodents, birds, mammals, snakes and most of the larger earthworm pests. Of course, screens and gratings placed at the top and bottom of the beds can also be effective, but you can never have too many lines of defence. A sheet of Mosquito netting draped over your bins would eliminate most flying pests and is little hassle to use.

How to Deal with Worm Composting Pests

All of the following creatures pose a threat to earthworms: ants, mites, slugs, raccoons, springtails, rats, moles, amphibians, reptiles, gophers, certain beetle larvae, maggots, and a variety of other insects. Fortunately, most of these villains can be neutralised by properly constructed bins, screening, or – most importantly – good worm bed management. Nonetheless, we'll take a closer look at some our beloved worm's greatest enemies and what can be done about them.

Ants

Watch out for ants as they can wreck your beds in a matter of days and therefore require immediate action. Ants are attracted to the feed, so don't spill any near your bins and clear away any old spillage as soon as it is spotted. If your bin isn't too big and has legs, another way to keep ants out is to put each of your bin's legs in a dish of water – alternatively, most of the garden centres sell ant goo – a sticky substance that is painted around the stems of rose bushes to trap ants. It is eco friendly as it doesn't contain any insectide poisons. If all else fails and the ant invasion has already become serious, you can dust the area around your beds with pyrethrum dust or douse the ant nest and the trails leading to your bin with a granular insecticide, or use commercially available ant traps, which contain slow release poisons that the ants take with them back into their nests. Please be sure not to use any insecticide on the actual worm bed soil or you will kill your worms. If ants are already established inside the beds soak the section they are in and they will usually go away. These tiny creatures do not actually harm your worms, but are unsightly and do compete with the worms for available food. Most worm beds usually contain several species of mites (the most important for, our purposes, being the earthworm mite), which pose no real threat to the worms unless their population spirals too high – this usually happens as a result of poor bed management. Earthworm mites are small and are usually brown, reddish or somewhere in-between. They tend to concentrate near the edges and surfaces of the worm beds and around clusters of feed. They are not known for attacking the earthworms but do eat the earthworms feed. When the mite population is too high the worms will burrow deep into the beds and not come to the surface to feed, which hampers worm reproduction and growth. High mite populations usually result from: –

 Over-feeding. Maintaining a proper feeding schedule (for example: one that ensures the feed is eaten in a few days) will prevent the feed from going off in the beds.

Feeding the earthworms meaty or wet feed. Large mite populations are often the result of using over moist garbage and vegetable refuse as feed. Adding the occasional soggy vegetable leftover probably won't cause a problem but don't make a habit of it.

Over-watering. A rule of thumb when watering is to keep the beds damp but not wet. Poor bed drainage can also facilitate a mite problem and make the beds less hospitable to worms. Ensure that there are adequate drainage holes at the bottom of your worm bin or housing.

Remember the same conditions that ensure high worm production will be less favourable to mites. If you find your worm farm overrun by mites, expose the beds to the sun for a few hours. Cut back on water and feed and then, every 1 to 3 days, add calcium carbonate. Another method is to over water the bed forcing the mites to the surface and then burning them with a blowtorch. Both of these methods though are only short-term remedies and eventually you will have to improve the conditions in your worm farm if you want to keep the mite population low.

Fruit Flies

These insects will be attracted by over ripe fruit and certain vegetable scraps. They lay their eggs in the decaying fruit, but are not really a major problem. Just make sure that you cover any fruit with some of the bedding. A jam jar that has a residue of sticky jam/ jello or marmalade smeared around its sides can be half filled with water and left beside the worm bin. The fruitflies are pretty stupid and get stuck to the jam or drowned in the water. I personally don't like to use commercial insectides, but fruit farmers often use ripe fruit bait that has been poisoned to attract and kill the fruitflies – I suppose its one step better than spraying the actual fruit that we are going to eat.

Blow Flies and House Flies

Excess flies buzzing around your worm bins or worm farms are usually the result of having used meat, greasy food waste, or pet faeces as feed. They spread disease and make life miserable for the worm farmer and his family. They can also result in maggots if the beds aren't properly sealed. If your farm is kept indoors or under some sort of shading – as it should be – then you can hang up some fly strips, which will draw them away from the farms. Again, a properly maintained worm farm will normally not stink and therefore not attract flies.

Black Soldier Fly

Latin Name: Hermetia illucens. It is a moot point as to whether this fly should actually be called a pest. It is a tropical fly, originally from the Americas, that has now spread around the world. The larvae of the fly are a type of small maggots, that feed exclusively on putrescent material. They are often found in worm farm bins, but although unsightly are not a real threat to the worms, as they do not attack them and may in fact complement the compost worm's activities, rather than compete with them for food. Like the vermiculture worms their faeces make excellent compost and the maggots are also useful as a high protein fish or poultry feed and may be used either live or dried, as a processed meal. They may also be used by the less squeamish for fish bait. They can best be kept out of the worm farm bins, by not using meat and fatty waste and by keeping the moisture on the dry side, and making sure that there is a good cover of bedding material over the feeding area.

These remarkable creatures, unlike the common housefly, do not spread bacteria or disease – in fact the larvae ingest potentially pathogenic material and disease-causing organisms and thus render them harmless. Moreover black soldier flies exude an odour, which positively discourages houseflies and certain other flying pests. When the larvae reach maturity they leave the feeding area to pupate, preferably in a shady bush or tree. After turning into an adult fly, the female lives a further 5-8 days and produces almost 1000 eggs. The adult fly is nocturnal and characterised by very fast and rather clumsy flight. It has no mouth and cannot bite or sting. There is a growing interest in using Black Soldier Fly for commercial processing of sewage and agricultural waste. Some hobbyists have been experimenting with the Black Soldier Fly, as an alternative to vermiculture, for for private composting/ waste disposal. For the same size of container it is said that a well stocked colony of Black Soldier Fly would be able to process waste material very much faster than a comparable sized worm farm.

Springtails

These wingless oblong insects live on decaying and sometimes living plant matter and are a subclass Apterygota. You can recognise them because they jump when disturbed and can turn a worm bed surface white if the population is large enough. Although they have on occasion been observed to eat dead or weak worms, they are primarily a nuisance because they eat the worm's food and can, when the populations are big enough, drive the worms deep into the beds and keep them from coming to the surface to feed. One deals with them the same way one deals with mites.

The Worm Dictionary and Vermiculture Reference Center

Worm Terms

This page is to be a combination of vermi-dictionary, glossary and mini worm encyclopaedia. It is intended to be the first point of reference for all vermiculture terms and definitions and as such will probably never be completed, as it is to be updated regularly. As you can see, the cupboard is still a little bare, but be patient – it is filling up nicely.

A B C D E F G H I J K L M N O P Q R S T U V W X Y Z

A

Acidity:

The optimum condition for a worm farm is to have a balanced or neutral pH of 7. A common problem is that over a period, the environment becomes acidic, with a low pH value, especially if the environment is too wet. The worm farmer should always avoid feeding with acidic (sour) fruits such as citrus and pineapple. The needles of coniferous trees such as pines are also acidic. Mites are believed to be promoted by acidic conditions (pH less than 7).

African Nightcrawlers:

Latin name: Eudrilus Eugeniae. These worms are much larger than Eisenia Fetida (Red Wigglers) and are commonly over six inches long. Good compost worms and great for fishing, because of their size and as they are lively on the hook and have a firm skin. They prefer temperatures of around 75ºF- 85ºF , but can tolerate 45ºF- 90ºF, cannot tolerate extreme cold and dislike disruption of environment and handling. Weight: 175 to 200 worms per pound.

Anecic Worm Species.

These are the deep burrowing earthworms, such as Lumbricus Terrestris. They are not suitable for vermiculture as they are not compost worms – however in nature, they are extremely beneficial in improving the overall soil condition. Usually greyish brown, they are excellent diggers and their burrows, are semi permanent and may extend to six feet below the surface – these burrows are lined with mucus and help aerate the soil and improve water retention. The worms feed near the surface and pull down organic material deep into the ground, putting the nutriments in the worm castings exactly where the plant roots can best use them. They block the mouths of their burrows with plugs of soil and castings, to keep out unwanted pests.

B

Bait worms:

See Fishing Bait.

Bedding:

This is the medium placed at the bottom of the layers of the worm farm to provide a habitat for the worms. It may be of coconut coir, carpet undefelt, partly composted straw, crumpled cardboard or torn up paper – in fact almost any fibrous organic material. The food (kitchen waste etc.) is placed on top of the bedding, which will gradually be ingested and converted to worm casts (the worm compost). It is usual to cover the food and casts with a further layer of bedding material to exclude light and keep out flying pests. As compost worms need a good supply of oxygen, the bedding must be gently fluffed up from time to time to prevent it becoming dense and matted.

Bins:

This is the actual container that contains the vermiculture activity, they may be made of plastic , timber or any other non toxic material. A worm farm may consist single or multiple bins. The most popular type of worm farms today would consist of several levels of nesting plastic bins. These worm farms are sold at retailers, garden nurseries or on-line merchantizers and are designed to make them easy to handle. They usually have legs and a tap for the worm tea incorporated in the bottom bin. The plastic bins they use are neat and easy to keep clean and secure from pests. For the DIY enthusiast, household plastic storage containers, bought from the local hardware store are easy to adapt for setting up a DIY worm farm as they are robust, cheap and usually supplied with a snap-fit lid. As they are tapered for empty stacking, they can be nested to set up a multi layered worm composter. The popular sizes are 45 litre (10 gallon) or 6o litre (15 gallon). For worm farms, they should preferably be black and never be clear, as compost worms hate light.

Black Soldier Fly:

Latin Name: Hermetia illucens. This is a tropical fly originally from the Americas, that has now spread around the world. The larvae of the fly are a type of small maggots, that feed exclusively on putrescent material. They are often found in worm farm bins, but although unsightly are not a real threat to the worms, as they do not attack them and may in fact complement the compost worm's activities, rather than compete with them for food. Like the worms their faeces makes excellent compost and the maggots are useful as a high protein fish or poultry feed and may be used either live or dried, as a processed meal. These remarkable creatures, unlike the common housefly, do not spread bacteria or disease – in fact the larvae ingest potentially pathogenic material and disease-causing organisms and thus render them harmless. Moreover black soldier flies exude an odour, which positively discourages houseflies and certain other flying pests. When the larvae reach maturity they leave the feeding area to pupate, preferably in a shady bush or tree. After turning into an adult fly, the female lives a further 5-8 days and produces almost 1000 eggs. The adult fly is nocturnal and characterised by very fast and rather clumsy flight. It has no mouth and cannot bite or sting. There is a growing interest in using Black Soldier Fly for commercial processing of sewage waste and even for private composting/ waste disposal.

C

Casings:

See Cocoons.

Clitellum Band:

A band around the worms upper body, of lighter coloured flesh, found a centimetre or so behind the front of sexually mature worms. It is from the clitellum that the egg casings (cocoons) form as rings before slipping off the worm's bady, containing both the eggs and sperm from the earlier mating.

Cocoons:

The worm cocoons or egg casings start to be seen in the bedding as soon as the worms become sexually mature at the age of about 8 to 10 weeks after hatching. Shortly after hermaphrodite mating has taken place, the light coloured clitellum band near the front of the worm secretes a mucous cocoon ring, which slips off the worm. As it slips free, it receives the worm's eggs, together with the sperm from the other worm. The cocoon or casing curls into a ball around the eggs. The cocoons are about the size of bb gun pellets or grape pips and are pale yellowish brown to grey at first, becoming darker and mahogany coloured as they get near to hatching at about seven weeks. Each casing yields several tiny pale coloured baby worms.

D

Detritivorous:

Literally "eater of trash" – Detritus is the polite Latin name for plant trash and animal dung that is found lying on the surface of the soil. Detritivorous creatures, such as wood lice, black soldier fly and notably the epigeic (composting) worm species, such as Eisenia fetidae (red wigglers) all eat decomposing organic matter and, in association with beneficial bacteria, convert this waste into useful plant nutriments.

E

Egg Casings:

See Cocoons.

Eisenia Andreia:

Usually called the Tiger Worm, because of alternate bands of darker and lighter red colour.

Often confused with Eisenia Fetida (Foetida) and to make things worse they are also known as Red Worms. Like Fetidae They are quick breeders and productive in vermicomposting and good fishing worms. They are between 2 to 3 inches long and weigh in at 900 to 1000 worms per pound. They are found throughout the world and as such are no threat to the environment if they escape. Temp range – Extremes: 38ºF-88ºF/Optimum 70ºF -80ºF.

Eisenia Fetida (Foetida):

Commonest compost worm used in worm farming and easy to obtain. Usually called Red Wigglers, but also known as Red Worms, Red Wrigglers, Compost Worms, Manure Worms and Brandling Worms. They got their name of red wiggler because as fishing worms as they are active on the hook and stay alive in water for some time, although they are a bit small for this purpose. They are between 2 to 3 inches long and weigh in at 900 to 1000 worms per pound. They are quick breeders and productive in vermicomposting. They are found throughout the world and as such are no threat to the environment if they escape. Temp range – Extremes: 38ºF-88ºF / Optimum 70ºF -80ºF

Eisenia Hortensis:

Common name: European Nightcrawler also commonly called Redworm, it is much bigger than Eisenia Fetida (foetida). It is a quick breeder and a good composter (makes plenty of castings). Much sought after for fishing bait, as it can tolerate near freezing water and is one of the few "earthworms" suitable for salt water fishing. These worms can grow up to 7 inches in length, but usually are between 3 to 4 inches. 300 to 400 worms per pound.

Endogeic Worm Species:

These are the upper soil worm species of earthworm, and are geophages as they feed on humified soil with high organic content such as found around grass roots. They make temporary burrows, which become filled with their castings and this brings nutriments to plant roots and their burrowing improves the aeration and moisture retention.

Epigeic Worm Species:

These are the surface dwellers ,such as compost worms. They include the smaller red/ brown worm species, that are to be found naturally just below the surface in rotting leaves, dung heaps and other plant litter. They are called detritivourous in that they eat detritus (waste material). They can handle cycles of variable moisture, but will not survive long in soil, unless it has been well loosened and mixed with compost or manure as they are poor diggers and need plenty of organic matter around them and must have good aeration. Their natural home is within the plant trash itself .

Eudrilus Eugeniae:

Common name: African Nightcrawlers. These worms are much larger than Eisenia Fetida (Red Wigglers) and are commonly over six inches long. Good compost worms and great for fishing, because of their size and as they are lively on the hook and have a firm skin. They prefer temperatures of around 75ºF- 85ºF , but can tolerate 45ºF- 90ºF, cannot tolerate extreme cold and dislike disruption of environment and handling. Weight: 175 to 200 worms per pound.

European Nightcrawler:

Latin name: Eisenia Hortensis also commonly called Belgian worm and Carolina crawler and sometimes Redworm, it is much bigger than

Eisenia Fetida (foetida). It is a quick breeder and a good composter (makes plenty of castings). Much sought after for fishing bait as it can tolerate near freezing water and is one of the few "earthworms" suitable for salt water fishing. These worms can grow up to 7 inches in length, but usually are between 3 to 4 inches long and weigh in at 300 to 400 worms per pound.

F

Fishing Bait:

There is a large industry involved in producing bait worms. The common earthworm, such as Lumbricus terrestris, although can be dug up easily enough, is not suitable for commercial production as are the red compost worms as it is a soil dwelling species and is not productive in worm farming. Red wigglers, (Eisenia fetida) are commonly used , but, although lively on the hook are rather on the small side. Lumbricus Rubellus (also called red wriggler or red worm) are larger, more robust and can even be used for saltwater fishing. Other popular bait worms are Eisenia Hortensis – the European nightcrawler, which can come up to seven inches and tolerate near freezing conditions and the large African nightcrawler Eudrilus Eugeniae for warmer waters.

G

Geophages:

Fauna, such as earthworms, termites and other small species that consume humus soil which has a degree of organic content.

H

Hermaphrodite:

Having both male and female genitalia (from greek). Worms such as earthworms and compost worms are hermaphrodite and in a single "sexual encounter" each partner produces both eggs and sperm.

Hermetia illucens:

See Black Soldier Fly.

I

Indian Blue Worm:

Latin name : Perionyx excavatus. This species has a distinctive iridescent blue sheen to its skin. It is a tropical worm and does not tolerate cold or much handling or environmental disruption. Although small, it is suitable for vermiculture as it is a prolific breeder and matures quickly. It has one major drawback though – it is known for staging mass escapes from the worm farm, for no apparent reason and is somewhat unpopular for this reason. Temperature range – Extremes: 45ºF – 90ºF / Optimum 70ºF – 80ºF.

J

No 'J' yet:

Still looking.

K

No 'K' yet:

Still looking.

L

Lime:

Agriculural lime is Calcium Carbonate CaCo3 (usually finely ground up limestone or chalk). Ground up sea shells, egshells or even marble all contain Calcium Carbonate and are safe to use with your worms in small quantities. Never use quick lime, Calcium Oxide (CaO), which is highly corrosive with a pH of 14, nor should you use the milder, but still highly alkaline Calcium Hydroxide Ca(OH)2 – known as builders (slaked) lime.

Lumbricus rubellus:

A species of European worm, the driftworm, also known as Red wriggler. It is actually an burrowing earthworm and not a true compost worm, but in nature is Endogeic and feeds close to the surface. It is a large worm of average length 4 inches and is commonly used in vermiculture, as it is very productive at cooler temperatures. The optimum temperature is around 50ºF and it only stops breeding around 40º. Rubellus is also attractive as a bait worm as it is large, lively, robust and is even suitable for salt water fishing. However there is real concern that Lumbricus rubellus, as an exotic, could become a problem invasive species in North America and there are claims that it is spreading into the northern woods and causing damage to native forests. This is because it tolerates lower temperatures and wetter conditions than most compost worms. It causes damage by breaking down the plant subterranean trash that protects the surface roots of trees. Because it can burrow deeply, it can overwinter when the surface becomes frozen, unlike most compost worms such as Eisenia fetida. So before you start your worm composting – it is important that you check local requirements and choose the right worms for your area and never throw unused bait into the forest.

Lumbricus terrestris:

Common earthworm species, sometimes called nightcrawlers . They are not suitable for vermiculture as they are a deep burrowing species (Anecic). Their burrows, are semi permanent and may extend to six feet below the surface – these burrows are lined with mucus and help aerate the soil and improve water retention.

M

Microbes:

Tiny living organisms, only visible with a microscope, such as bacteria, protozoa, some fungi and viruses. They are also known as micro-organism. They may be pathogenic, causing disease and damage to plants, animals and humans (pathogenic microbes are commonly called germs) or beneficial such as the bacteria that promote yeast fermentation and the aerobic microbes that break down decomposing waste and improve soil fertility. Some beneficial micro-organisms actively attack pathogenic bacteria and decrease the risk of disease. Micro-organisms associated with vermiculture are especially good for the garden, as working inside the worm's gut they promote the digestion process and breakdown complex chemical compounds into simpler elements that are more easily utilised by plants, when excreted. By "fixing" the essential elements into simple compounds, minerals, hormones and enzymes they hugely increase the accessible plant nutrient load in the soil (up to 20 times) and as they remain active in the faeces (castings) they are also available to "innoculate" the soil against disease. In the aerobic breakdown of organic matter by bacteria some carbon dioxide (CO_2) is released, but this is far less damaging to the atmosphere than the methane (CH_4) that would be released otherwise by anerobic decomposition. On the plus side, other bacteria also create conditions for fixing some of this CO_2 in the soil, for use by plants..

Micro-organisms:

N

No 'N' yet:

Still looking – Why not send us your 'B' ?.

O

No 'O' yet:

Still looking – Why not send us your 'O' ?.

P

Perionyx excavatus:

Common name : Indian blue worm. This species has a distinctive iridescent blue sheen to its skin. It is a tropical worm and does not tolerate cold or much handling or environmental disruption.

Although small, it is suitable for vermiculture as it is a prolific breeder and matures quickly. It has one major drawback though – it is known for staging mass escapes from the worm farm, for no apparent reason and is somewhat unpopular for this reason. Temperature range – Extremes: 45ºF – 90ºF / Optimum 70ºF – 80ºF.

pH Value:

pH: A scientific scale from 1 – 14 that is used as a measure of the acidity / alkalinity of a substance. A pH rating of 7 indicates a neutral or balanced rating, that is neither acidic nor alkaline. The lower the pH number, the more acidic the substance will be and vice versa, the higher the pH number – the more alkaline. pH meters or testing kits are available from garden suppliers (for soil testing).

Q

No 'Q' yet:

Still looking – Why not send us your 'Q' ?.

R

Red Wiggler:

Latin name: Eisenia foetida (fetida) This is the commonest compost worm used in worm farming and is easy to obtain. Also known as Red Worms, Red Wrigglers, Compost Worms, Manure Worms and Brandling Worms. They got their name of red wiggler because as fishing worms as they are active on the hook and stay alive in water for some time. They are between 2 to 3 inches long They are found throughout the world and as such are no threat to the environment if they escape. Temperature tolerance – Extremes: 38ºF – 88ºF / Optimum 70ºF – 80ºF. They weigh in at 900 to 1000 per pound. They are quick breeders and can tolerate some handling and disturbance to their environment and they are productive in vermicomposting.

Red Worm:

Several compost worms are known by this name, although due to their popularity for vermiculture, it is usually applied to Eisenia fetida (foetida) a.k.a – Red Wrigglers, Compost Worms, Manure Worms and Brandling Worms and the name is also applied to Eisenia foetida's close cousin Eisenia andreia (andreii) a.k.a Tiger worms, as both worm species are virtually indistinguishable and have similar environmental preferences. They are both between 2 to 3 inches long and weigh in at 900 to 1000 worms per pound. They are quick breeders and productive in vermicomposting. They are tolerant to handling and disruption of their environment. They are both found throughout the world and as such are no threat to the environment if they escape. Temperature tolerance – Extremes: 38ºF – 88ºF / Optimum 70ºF – 80ºF.

S

Stacked Bin Worm Farm:

The stacked bin worm farm usually consists of a stack of nesting plastic containers to hold the worms, either circular or rectangular shaped. The lowest container usually has a tap for the leachate (worm tea) and possibly legs. The upper containers of the stack have perforated bases and small air holes for ventilation and there would be a lid to close the topbin. For a small operation 3 layers would be sufficient, but stacking towers up to 7 levels are known for farms at restaurants and hotels. The Stacked bin system can be bought "ready-to-go" from a merchant or made up inexpensively by the DIY enthusiast. Other materials such as wood or fibre board, could also be used.

Stacked Tire Worm Farm:

An old way of breeding worms for fishing, especially popular on agricultural farms and smallholdings. Used tires are stacked up one above the other to make a worm farm. Can also be used for general vermiculture, as a very cheap alternative for the stacked bin system – especially in poor countries. The compost is removed, by pulling out the lower tire from the stack (usually with the help of crowbars, due to the weight of the tires).

T

Tiger Worms:

Latin name: Eisenia andreia gets the name Tiger Worm, because of alternate bands of darker and lighter red colour. Often confused with Eisenia foetida (fetida) and to make things worse it is also known as Red Worm. Like fetida, they are quick breeders and productive in vermicomposting and good fishing worms. They are between 2 to 3 inches long and weigh in at 900 to 1000 worms per pound. They are quick breeders and productive in vermicomposting. They are tolerant of handling and disruption of their environment. They are found throughout the world and as such are no threat to the environment if they escape. Temperature tolerance – Extremes: 38ºF – 88ºF / Optimum 70ºF – 80ºF.

U

No 'U' yet:

Still looking – Why not send us your 'U' ?.

V

Vermi:

Latin word for worm, as used as a prefix in vermiculture, vermicompost, vermitea etc..

Vermicompost:

This is the highly nutritious plant food that is the product of worm farming. It is actually the worm castings or faecal matter remaining after waste organic material such as kitchen scraps have passed through the gut of the compost worms. It has typically around twenty times the nutritional load of good quality garden soil and contains many minerals and race elements necessary for healthy plant growth. Vermicompost is also accompanied with a load of beneficial bacteria, which promotes healthy soil conditions and has the ability to remove pathogens and various unwanted organisms from the garden soil.

Vermiculture:

A slightly broader term than worm farming, in that it that places equal weight on three aspects of the process. a) the environmentally important aim of getting rid of waste products; b) on generating useful plant compost from the worm castings and c) on the breeding of worms themselves. They may be used as fishing bait or even as a source of high protein poultry, fish or animal fodder.

Vermitea:

This is the popular term for the dark highly nutritious leachate or liquid fertilizer associated with the worm composting process. It is excreted by the worms, together with the worm castings in the process of breaking down organic material that has some degree of moisture content. It is usually tapped off from the underside of the worm farm into a watering can then diluted for pouring directly around the roots of garden plants. Vermiteais is also accompanied with a load of beneficial bacteria, which promotes healthy soil conditions and has the ability to remove pathogens and various unwanted organisms from the garden soil.

W

Worm Farm:

A worm farm is essentially a protected environment, set up for the nurturing of either earthworms, or more commonly, compost worms. The aim of the worm farm may be simply to produce worms for fishing bait, or it may have the wider aims of vermiculture, which are to produce good quality, highly nutritious plant compost (organic ferilizer) from waste organic matter, together with the worm production, thereby benefitting the environment, by replacing the need for inorganic fertilizers and cutting the environmental costs of waste collection to landfill.

X

No 'X' yet:

Still looking – Why not send us your 'X' ?.

Y

No 'Y' yet:

Still looking – Why not send us your 'Y' ?.

Z

No 'Z' yet:

Still looking – Why not send us your 'Z' ?.